spot

BIG BIRDS

OSTRICHES

by Lisa Amstutz

AMICUS | AMICUS INK

eye

legs

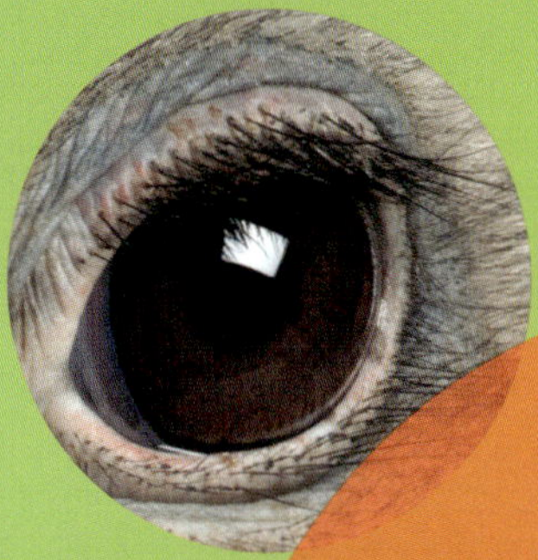

Look for these words and pictures as you read.

toes

wings

Look at that big bird.
It is taller than most people!
It is an ostrich.

See that big eye?
It has thick lashes.
They keep out sand.

eye

See the long legs?
Zoom! The bird runs.
No other bird runs faster.

legs

See those big toes?
Each foot has only two.
Other birds have three or more.

toes

wings

See the wings?
The bird cannot fly.
Wings help it turn.

Ostriches eat plants.
They eat bugs and lizards, too.

Ostriches lay the biggest eggs. The chicks are as big as grown chickens.

eye

legs

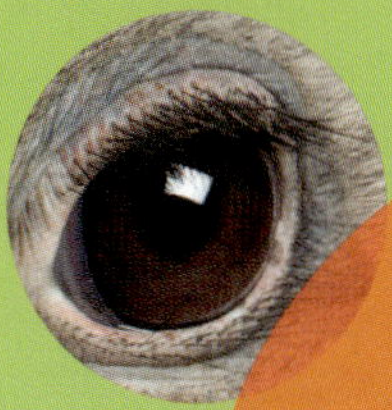

Did you find?

toes

wings

Spot is published by Amicus and Amicus Ink
P.O. Box 227, Mankato, MN 56002
www.amicuspublishing.us

Library of Congress Cataloging-in-Publication Data
Names: Amstutz, Lisa J., author.
Title: Ostriches / by Lisa Amstutz.
Description: Mankato, MN : Amicus, [2023] | Series: Big birds | Audience: Ages 4-7 | Audience: Grades K-1 | Summary: "The world's biggest bird, the ostrich, is showcased in this leveled reader about its habitat, body parts, and behavior. Includes a search-and-find feature to reinforce key vocabulary"-- Provided by publisher.
Identifiers: LCCN 2019060283 (print) | LCCN 2019060284 (ebook) | ISBN 9781645490999 (library binding) | ISBN 9781681526669 (paperback) | ISBN 9781645491415 (pdf)
Subjects: LCSH: Ostriches--Juvenile literature.
Classification: LCC QL696.S9 A47 2023 (print) | LCC QL696.S9 (ebook) | DDC 598.5/24--dc23
LC record available at https://lccn.loc.gov/2019060283
LC ebook record available at https://lccn.loc.gov/2019060284

Gillia Olson, editor
Deb Miner, series designer
Ciara Beitlich, book designer
Bridget Prehn, photo researcher

Photos by iStock/AaronAmat cover, 16; iStock/GlobalP 1, 4–5; iStock/001_Quadrixx 3; Alamy/Godong 6–7; iStock/LAByrne 8–9; Minden/Gerard Lacz 10–11; Alamy/Ciobaniuc Adrian Eugen 12–13; iStock/pilesasmiles 14–15